Pocket Guide to Coin-op Vending Machines

John Carini

4880 Lower Valley Road, Atglen, PA 19310 USA

Published by Schiffer Publishing Ltd.
4880 Lower Valley Road
Atglen, PA 19310
Phone: (610) 593-1777; Fax: (610) 593-2002
E-mail: Schifferbk@aol.com
Please visit our web site catalog at www.schifferbooks.com
We are always looking for people to write books on new and related subjects.
If you have an idea for a book, please contact us at the above address.

This book may be purchased from the publisher.
Include $3.95 for shipping.
Please try your bookstore first.
You may write for a free catalog.

In Europe, Schiffer books are distributed by
Bushwood Books
6 Marksbury Avenue
Kew Gardens
Surrey TW9 4JF England
Phone: 44 (0) 20 8392 8585
Fax: 44 (0) 20 8392 9876
E-mail: Bushwd@aol.com
Free postage in the UK. Europe: air mail at cost.

Designed by Joseph M. Riggio Jr.
Type set in Geometric 231Hv BT/Humanist 521 BT

ISBN: 0-7643-1658-3
Printed in China
1 2 3 4

Contents

Acknowledgments

When I first started collecting in 1987, I didn't know very much about coin-operated machines. But thanks to a number of special people, I've come a long way. Some of my knowledge comes from books, some from personal conversations with fellow collectors, and some from the machines I have purchased and sold throughout the years. In my early days of collecting, I often called and visited the late Marshall Larks (The Gumball King). Marshall was my mentor and personal friend. I remember driving down to his well stocked gumball warehouse in Chicago on many occasions in the late '80s.

My library contains several well used coin-op books from the late Dick Bueschel, friend, historian and author; and also from the late Bill Enes, author and collector. Dick Bueschel provided some of the earliest coin-op reference books, and his love of history was evident in his presentation style. Bill Enes provided encyclopedia style coin-op books, covering thousands of different coin-op vending machines. And lastly, I would like to thank my current collector friends with whom I have frequent conversations, including Hoyt Perkins, Paul Hindin, Jack Fruend, and Rich Brinkos.

Photo Credits

Thanks to Paul Hindin and Jack Fruend, who opened up their homes to me so I could photograph their extensive collections, to Rich Brinkos for providing the machines that were photographed for the cover, and to Chad Boekelheide, Hoyt Perkins, Doug Hess, Dan Sailer, Dan Wysinski, Ken Schaffer, and Jack Laquidara for sending in photos from their collection.

Introduction

It all started back in the spring of 1987 when my wife, Sandy, kept pestering me to accompany her on trips to the local flea markets. At the time, I couldn't see what attracted her to flea markets; I wasn't interested in the hours of endless walking and rummaging through someone else's junk. But she continued to drag me there anyway, explaining, "Just find something old that fascinates you, and you will get hooked too." And sure enough, at one flea market I came across a late 1940s double perk-up nickel breath pellet machine for $45. It was neat! At the time I didn't know if it was a good deal or not, but I knew I had found my collectible. Since then, I've bought and sold nearly 900 coin-op vending machines, and I'm always looking for that next great find.

I purchased this double perk-up nickel breath pellet machine in 1987 for $45. This is the first coin-op machine I ever bought.

Oh, about those flea markets I used to hate. You will now find me at every flea market within a 100 mile radius of Milwaukee on a regular basis. I'm glad that my wife and even my children, Nick (14) and Katie (12), enjoy collecting too because when we schedule our family vacations we can include visits to local antique shops and flea markets along with the standard local sightseeing. You know your whole family is a little antique nuts when you visit Orlando for a week and the kids suggest we set a day aside for antiquing!

Dragons Found in Wisconsin

One day in October of 1997 I received a postcard from an elderly man living in a nearby city who owned a coin-op machine. He had seen my advertisement in one of the area antiques newspapers and wanted to sell. The postcard he sent me read as follows:

> *I have a peanut machine (old). Our son took it apart to clean it up (this was 35 years ago) and never put it back together. To the best of my knowledge all of the parts are there. It should be worth $50.*

The city he lived in was about a 1-1/2 hour drive from here and at first I hesitated. Then my wife told me I should go, that it might be something worthwhile. I called his house and his wife answered. She described the machine as having a wood cabinet and a heavy metal front with two embossed dragons on it. The description was intriguing, so I scheduled a time to come over and take a look at it. From the description, I thought it would be a slot machine.

When I arrived, the man brought out a box of parts. As I looked them over, I realized this was a very old peanut machine, but I wasn't sure exactly what it was. The elderly couple who owned it knew it was worth more than the $50 they were asking, but indicated they wanted someone to put it together and take care of this machine that had been in their family for fifty years. I asked the man where he had originally gotten the machine, hoping to find out more. He said he picked it up about fifty years ago, real cheap, from a man who had a vending route in that area. The machine was a part of that route. He mentioned that at the time he picked this machine up (at the route operator's warehouse), there were quite a few more. He also said that the route operator had died many years ago, but I didn't let that stop me from asking if he could put me in touch with any of the route operator's relatives. Unfortunately, he didn't have any information on them.

I purchased this machine in a box, unassembled. After putting it back together I found a Freeport Dragon, in rough original condition.

As I looked around the house, I couldn't help but notice that the couple had a number of other antiques— wood tools, a 1930s slot machine, and a trade stimulator—all in excellent condition. I asked to purchase these items too, but the man said they weren't for sale. Satisfied with my one purchase, I left.

When I got the box of parts home, I was able to identify the machine using Bill Enes's book *Silent Salesmen Too* (see Bibliography) as a guide. After several days, when the parts were put together, I found a very rough but totally complete Freeport Dragon machine. What an incredible find! I was then able to start a basic restoration. My machine is in good condition, even though most Freeport Dragons haven't survived the years.

Every time I look at this machine, I can't help replaying in my mind the nursery rhyme about Humpty Dumpty and how all the king's horsemen couldn't put him back together again…and thinking just how fortunate I was that this man's son, even with all the parts there, couldn't put the Freeport Dragon back together again.

The Freeport Dragon after cleaning and simple restoration.

That's the story of how I found the most valuable piece in my collection. The following pages are devoted to the many vending machines you may encounter in your search. Some belong to me, and others belong to fellow collectors/friends who have supported my efforts. I hope you enjoy them as much as I have.

A Brief History of Coin-op Vending Machines

In the United States, vending machines have been in use since the 1880s. During the early years, machines were used to vend everything from candy, breath mints, pencils, perfume, and razor blades to toilet paper. Most were primarily constructed of wood or metal. Machines sprang up in train stations, general stores, smoke shops, and taverns. They became known as "silent salesmen," selling products twenty-four hours a day without a break.

The early 1900s vending machines were both ornate and complicated. Cast iron machines had claw feet and scroll designs. Glass globes kept the items inside fresh and allowed one to view the product before selection. However, mechanisms within these machines were complicated. Repair was often difficult and expensive.

This early gum vendor was used in the late 1890s.

This Leebold is an example of an early peanut vendor—note the beautifully ornate cast iron scrolls and glass globe.

During the 1920s and 1930s, vending machines were manufactured with steel construction. Porcelain finish over cast iron was popular. It provided a more durable finish and enhanced the overall appearance of the machine. Some very unique mechanical machines were produced during this period. One in particular, the Ro-bo, provided the user with entertainment as well as a vended product. When you dropped a penny in and pulled the handle, the gumball dropped down from the display area. The figurine then scooped it up and dropped it down the chute. Many other automated vending machines were produced throughout this time period as well. Because of these entertaining features, the machines are considered very collectible. This period, in particular, produced a number of machines that are today considered high end coin-op vending machine collectibles.

This 1920s Ro-bo gumball vendor provided entertainment along with the vended product.

We all know that during the Prohibition period of the 1920s and 1930s, alcohol was banned. Gambling was banned too, and because of that many slot machines were destroyed. Even some gumball machines were destroyed because they were viewed as gambling machines. The Hawkeye gumball machine shown here is an example of a gambling type coin-op vending machine. The vendor was set so that on every tenth pull a bell (built into the base) would ring and the customer would get his penny back along with a free gumball. These vendors were viewed by some state officials as gambling machines and therefore outlawed.

This Hawkeye vendor is a gambling gumball machine. The machine was set so that on every tenth pull, a bell would ring (drawing attention and creating excitement) and the customer got his penny back along with the gumball.

After World War II, machines changed to a simple aluminum or other metal or plastic construction. Mechanisms were simplified, making it easy for the route operators to repair or replace parts on the spot. If you grew up in the 1950s and 1960s you may remember seeing Ford, Northwestern '60, Toy 'n Joy, Victor, and Oak Acorn gumball machines in supermarkets and drugstores. Oak Manufacturing Co., in business since 1948, continues to manufacture the Oak Acorn gumball machine today, similar in function and design to the original.

Still easy to find and affordable, the machines of this era are the ones that most people start their collections with today.

The machines shown on pages 10-13 are still commonly found today at flea markets and antique shops. Shown in order are a Ford, Northwestern '60, Toy 'n Joy, Victor Halfback and Oak Acorn.

1950s Ford gumball vendor

1960s Northwestern gumball/peanut vendor—a common flea market find.

Late 1950s Toy 'n Joy
capsule vending machine.

Early 1950s Victor Halfback
gumball vendor—a great
starter machine.

Late 1940s Oak Acorn peanut vendor. A common flea market find, these vendors are still being produced today.

Depending on your age, you may remember taking a trip to the neighborhood candy store as a child and seeing gumball machines on the counter or by the door on stands. Outside of the local smoke shop you might find a scale. Drop in a penny and get your weight and fortune told. At the neighborhood gas station we would beg our parents for a dime to get a bottle of Coke. When it was time for the county fair, we'd hit the midway and play the arcade machines. We stood in line to work the claw machines, strength testers, and tractor push machines, then we waited in line to drop a coin in and have our fortunes told by a mechanical fortune-telling woman in a booth. As you will see, these fascinating coin-op machines can still be found today and make a fun and interesting collectible

Getting Started with Your Collection

The purpose of this book is to get you started on the hobby and to provide you with a price guide on coin-op vending machines. While not every vending machine is included in this book (there are thousands), the assortment chosen will provide you with a reasonable comparison should you find a vending machine not pictured. The book lists vending machines by category, in alphabetical order by common machine name. Once you get really serious about this hobby, there are several other excellent coin-op vending books available (see Bibliography) as well as information to be found on other coin-op machines such as slot machines, trade stimulators, pinball machines, and juke boxes.

Another way to connect with the hobby is to join the Coin-Op Collectors Association (C.O.C.A.). Membership information can be found on-line at www.coinopclub.org. This group offers a nice magazine style publication three times yearly and meets twice yearly on the Friday evening before the Chicagoland Antique Advertising, Slot-Machine & Jukebox Show. The meetings feature a guest speaker and a chance to network with other collectors.

Antique coin-op machines can be found at coin-op collectible shows across the country. Attending a show is a great way to make a decision on what you want to collect. These shows are held in major metropolitan areas, such as Pasadena, California; Indianapolis, Indiana; and the Arlington auction in Las Vegas, Nevada. The biggest one—and my personal favorite—is the Chicagoland show, which is held twice yearly in St. Charles, Illinois. More than 400 dealers from across the country make this one of the most incredible displays of coin-op machines and advertising. A walk through this show allows you to choose from the affordable common machines as well as the unusual and rare ones. You will find a huge variety of old coin-op machines here, including country store items, Coke machines, scales, slot machines, pinball machines, claw machines, trade stimulators, and old advertising—just about anything you can drop a coin into. Many of the biggest dealers and collectors will be at these shows, so it's also a great opportunity to meet and talk with them.

Other places to find antique coin-op machines include local flea markets, estate sales, auctions, and the Internet. Lastly, don't forget about local antique shops. I've picked up a number of bargains locally as well as on my business travels.

Something for Nothing

As children we are told not to expect something for nothing. But coin-op collectors know that's not always the case. Often when we purchase machines, they come without a key. As a matter of fact, the majority of the machines I purchase don't have them. I'm not sure why so many of them are missing; I guess somehow over the years they just get lost. While you might think this is a bad thing, I've found that it can actually be good. A machine sold without a key means the seller wasn't able to get into the machine to remove coins and old product often left behind.

I've purchased so many machines without keys over the years that I've created a little system. In the beginning, when the machine didn't come with a key, I visited the local locksmith. He helped me make a key for the machine and I always made an extra. Over time, that has turned into hundreds of extra keys. Now when I pick up another machine without a key (and before I visit the locksmith), I try my extra set of keys. It may take several minutes, but about half the time I find a key that works.

Once you get the machine open, you can sometimes find valuable treasures. In the coin box I have found old coins including wheat pennies, buffalo nickels, and mercury dimes. In the stamp machines I've found old stamps. My Roi-Tan cigar vendor (see photo) came complete with two full dispenser boxes of cigars. Sometimes these "extras" are worth even more than the machine itself!

This 1950s cigar vendor came complete with two boxes of vintage Roi-Tan brand cigars.

A fellow collector, Paul Hindin, related this story of old chewing gum he found in an antique coin-op gum vendor he purchased:

A fellow collector and I got a lead on a National Colgan's Taffy Tolu gum vendor from a Chicago dealer who had found it in an old barn. We decided rather than try to outbid each other, we would make a fair bid and purchase the machine together. Pleased with our $2,000 purchase, I took it home and opened it up to clean it. I was pleasantly surprised to find seventeen sticks of Colgan's Taffy Tolu Chewing Gum inside. In the past I had found other sticks of chewing gum, and they sometimes brought as much as $35 per stick. However, this gum was a mystery to me. So I called up the most knowledgeable man in the industry, Bill Enes. When I told Bill what I had found there was a long pause on the phone. He asked "Are you sure?" Bill then told me he wasn't certain of the value, but suggested I not let them go individually for less than $350 per stick—then offered to buy all seventeen sticks! Bill provided me with the names of some serious gum collectors to contact if I was interested in selling. There are about a half dozen serious gum collectors here in the U.S., and about two dozen not-as-serious collectors. Each collector I called couldn't wait to get their hands on a few sticks of this mystery gum, as none of them had this brand in their collection. In the end, my partner and I sold thir-teen sticks of the gum for $300 to $350 each, making a $4,000 profit without even selling the machine!

While not every purchase will lead to such a lucrative find, most coin-op dealers can tell you at least one story about getting something for nothing.

This gum vendor came with quite a surprise for the new owner—seventeen sticks of Taffy Tolu Chewing Gum valued at $300 to $350 per stick.

Restoration

As a vending machine collector, I keep my eyes open all the time. I also have to watch my budget because prices are going up and machines are getting harder to find. But if you pay attention to what you are looking for, and keep your mind on how you can bring some of these machines back to shape, little by little you can build up quite a collection. I do it all the time.

When you realize that many of these coin-op vending machines are seventy years old or older, it's easy to understand why so many are found with broken globes, missing parts, peanut damage, or in generally poor condition. While many collectors will pass on these machines, I use their poor condition as a bargaining advantage and often pick them up. Finding original globes or other replacement parts is difficult, however. Most experienced collectors know that in addition to looking for coin-op vending machines they need to keep their eyes open for old globes, chutes, and other original machine parts to use for future restoration projects. I once picked up a 1908 Climax peanut vendor missing the mechanism for $135, knowing that sooner or later I would be able to complete the machine, making it worth almost $1500. Three years after I purchased this machine, I was scanning the coin-op items on eBay and found a listing for a Climax mechanism. I was so excited to finally be able to complete my machine. Unfortunately, after a fierce bidding battle with another individual, I came up $1 short!

Peanut vending machines can come with a lot of problems. Often when the machines were taken out of circulation, they were left filled with old peanuts. Over time, the salt and oil from the peanuts will corrode and freeze up the mechanism of the machine. Probably half of the machines you find will need a little elbow grease to make them work and look good again. Here are a few things you can do.

For a mechanism that is frozen, totally submerge the machine in a big bucket of water. Add a liquid dishwashing detergent (one that says it will cut grease) to the water, and let sit for a couple of days. This helps to break up the salt and oils. Then remove the machine and lightly tap on the turn handle with a rubber mallet. If the machine is stubborn, repeat for up to a month. This is usually all that is necessary to restore the mechanism to working condition. Once the mechanism is free, take the machine apart and clean all the parts. If springs are broken, replacements can be purchased right at your local hardware store.

I purchased this Climax for $135 (missing the mechanism), knowing the globe alone was worth more than the purchase price.

For cracks in the vending wheel or the machine itself, your local welding shop can help. If the crack is minor, you can use body filler. If your machine comes without a key, take it to your local hardware or locksmith to have a new key made. If you need a new glass globe, a new metal center rod, springs, or other parts, you can contact me at jscarini@execpc.com and I will be glad to provide you with information on locating these parts.

One way you can totally restore your machine is through the use of a paint stripper. But an even better and cleaner method is to use a sandblaster. You can purchase a sandblaster yourself or you can take the machine somewhere to have this service performed. This strips the machine of its original paint and allows you to repaint the machine in any color and make it took like brand new.* When I repaint, I try to keep the paint in the original color. I like Hammered Finish™ spray paint for a number of reasons. It is available in a multitude of colors, you don't need a primer, and it hides imperfections in the metal. The machine is left with a clean new finish that resists rust and chipping.

*If the paint is still in good condition, however, I recommend you do not repaint it. A machine in good original condition is usually worth more, and is more desirable.

I purchased this Victor Model V peanut vendor for $8 at a local flea market.

The machine's mechanism was frozen so I soaked the machine in water and added liquid dishwashing soap until I was able to free it. Then I took the machine apart to clean and sandblast the pieces. Green Hammered Finish™ spray paint provides a nice looking finish, along with a new plastic front and decal. Restored value, $75.

I purchased this Regal at a flea market for $40.

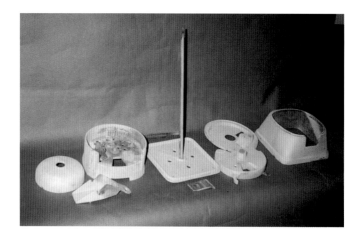

After disassembling the machine, I took the parts to be sandblasted.

I painted the chute, turn handle, and center rod using Hammered Finish™ silver spray paint.

I painted the balance of the parts using Hammered Finish™ blue spray paint.

I reassembled the machine and added a replacement globe and decal. Restored value, $100.

Reproductions

As with most other antiques and collectibles, you need to keep your eyes open for reproductions. Reproduction machines and vintage machines with reproduction parts are everywhere. Reproduction machines range from the inexpensive line of colorful Carousel* gumball machines you find at your local department stores to more expensive reproductions of some of the very old, hard to find novelty machines like the Smilin' Sam from Alabam'. One obvious but effective way to determine if the machine is a reproduction is to look at the casting to see what names or numbers are stamped. For example, the reproduction Baby Grand pictured below is marked Olde Tyme Reproductions Inc. on the front casing. The Columbia Carousel, pictured next to an original Columbus Model A, is marked Carousel® on the base of the machine. It's a bit harder to tell if the Smilin' Sam from Alabam' is a reproduction. It is currently being reproduced in aluminum and sells for around $650 (a fraction of the cost of the cast iron original).

The Canadian reproduction on the left is available for $40, while the original Baby Grand Deluxe on the right is valued at $75.

The Old Columbia reproduction on the left is available today for $59 from your local department store. The original Columbus Model A on the right is valued at $300.

* Carousel®, a division of Ford Gum & Machine Co, Inc.

These two aluminum reproductions of Smilin' Sam From Alabam' are available for $675 each, while the original cast iron version is valued at over $3000 (cast iron on left).

Reproduction parts are also available, which can be a good thing since so many machines are found with broken globes or missing parts. Some dealers have taken the time to create molds to reproduce the old style glass globes. In other cases, dealers have found the original old mold so they can create new, old stock. Sometimes dealers have parts re-cast from the original at their local foundry. It's not uncommon to find a vintage machine with some reproduction parts. That's why it is important to talk to the dealer and establish the originality of the machine. While still very desirable, these machines should be priced accordingly.

Pricing

Because machine conditions vary greatly, I've indicated a price range based on (1) a machine in original, fair to good working condition and (2) a machine without the vintage barrel locks found on pre-1940 machines (these original padlocks are collectible in and of themselves and can add $50 to the price of the machine). Any deviations from these criteria will be indicated next to the prices. Remember, machines you find may be missing parts, damaged, or in otherwise poor condition. The price you pay for these machines will depend on many factors, including whether or not you can find replacement or reproduction parts and whether the machine can ever be restored. It's important to talk to the dealer to establish this and price accordingly. Also, different regions of the country have slight variations on the pricing. For example, a Northwestern '33 machine is easy to find in the Midwest, but would be harder to find in the Northwest, making it slightly more expensive there. The word rare is used to describe machines where less than six are known to exist. I have not placed actual values on these, but if you could find one of them to purchase it would typically run in the thousands.

The Vending Machines
Aspirin

BLACKHAWK - early 1930s aspirin vendor. This six column nickel vendor is constructed of metal with a front glass display window. $400-$600.

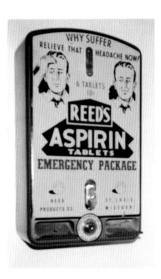

REED'S - 1940s aspirin vendor. This wall mounted vendor was constructed of sheet metal with advertising style graphics on the front. There were several similar versions of this machine made to vend other products. $300-$500.

Candy/Gum/Peanuts

ABBEY CASH TRAY - late 1940s peanut vendor. This small countertop machine was attached to a tray. It is aluminum construction with a round glass globe. The photo on the left shows the original globe and decal, while the photo below has a replacement (lantern style) globe and decal. $75-$110.

ADAMS PEPSIN GUM - early 1900s gum vendor. Beautiful embossed cast iron machine has a glass front viewing window. Embossed graphics read PURE GUM. Rare.

AD-LEE E-Z - 1930s gambling gumball vendor. This cast iron machine has a marquee bolted to the top cap (marquees pictured are reproduction). The gum you find in this vendor has its center drilled out and small slips of paper placed inside. The customer removed the slip of paper from the gumball and matched the numbers to the marquee at the top of the machine to determine the prize won. Prizes were redeemed from the store clerk. These two photos show the same machine with two different gambling/prize marquees. Original $1000-$1200; with reproduction marquee (as shown) $650-$850.

ADVANCE 1-2-3 - 1920s gumball vendor. The front of this machine has a special window allowing customers to view the specific color/gumball they would receive before inserting their coin. Hard to find. $600-$750

ADVANCE - 1920s gumball vendor. This cast iron machine has a small football shaped globe and protruding mecha-nism. $200-$350.

ADVANCE -
1930s gumball
vendor. Cast iron
machine with a
uniquely shaped
globe. $700-$800.

ADVANCE MODEL D -
1920s gumball vendor.
Machine is steel construction with a protruding
mechanism. $175-$275.

ADVANCE - 1920s candy/gumball vendor. This unique metal wall mounted machine was often found mounted to the back of movie theater seats. $350-$500.

ADVANCE MODEL D - 1920s gumball vendor. Steel construction vendor has a chrome top and mechanism. Good starter machine as they are still fairly common and not too expensive. $100-$200.

ADVANCE MODEL 11 - 1920s peanut vendor. Nicknamed the Big Mouth because of the large built-in tray, this tall vendor is constructed of steel. $175-$275.

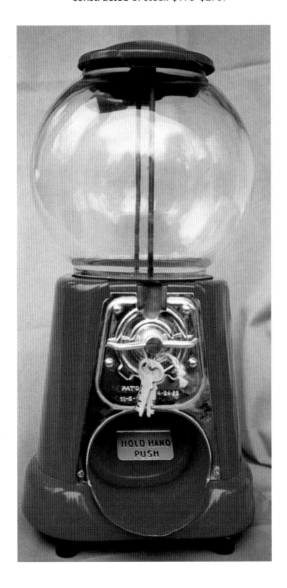

ADVANCE MODEL C - 1920s gumball vendor.
This tall, slim wall mounted gum vendor came in
a number of versions, dispensing gum or candy
in both 1¢ and 5¢ versions. $50-$150.

AJAX TRIPLE HOT NUT VENDOR - late 1940s hot nut vendor. This vendor has a polished aluminum construction and is typically found mounted on a cast iron stand. $300-$500.

ARNOLD'S - early 1900s stick gum
vendor. Small machine has a wood
case and ornate metal front with
advertising window and nicely
painted side graphics. Rare.

ASCO HOT NUT VENDOR - 1940s peanut vendor. This 5¢ polished aluminum hot nut vendor came with an optional cup holder. Machine above right is in original condition, machine below was repaired and repainted, most likely due to peanut damage. Without cup holder $100-$200; with cup holder $150-$250.

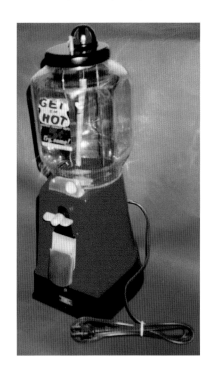

ATLAS ACE - 1930s peanut vendor. This Art Deco style machine is constructed of aluminum. Machine is pictured with a replacement bell shaped globe; if found in original condition it would have a cylinder shaped globe. $100-$175.

ATLAS BANTAM - late 1940s peanut vendor. Small counter machine with an attached chrome tray base and side coin entry/turn handle. Aluminum construction with ribbed glass globe. $100-$175. (Center left and bottom right)

ATLAS MASTER - 1950s gumball/ peanut vendor. An interesting feature of this machine is that the mechanism is designed to take either a penny or a nickel. The customer dropped in a penny for one turn or a nickel for five turns of the handle. The machine is constructed of aluminum and has a heavy, square glass globe. $35-$75.

ATLAS MIDGET - 1950s gumball vendor. This small counter machine can be found with or without a chrome tray base. Sleek design with a tall, slim glass globe. Without tray $75-$150; with chrome tray $125-$200.

AUTOSALES VENDOR -
Late 1890s candy/gum
vendor. This wall mounted
vendor was quite large and
heavy. Constructed of wood
and metal, this vendor is very
hard to find. $1400-$1800.

BANTAM BEEHIVE - 1940s gumball vendor. A small countertop tray vendor constructed of polished aluminum. $400-$600.

BEAVER - 1970s gumball vendor. Canadian machine with aluminum construction and a high capacity plastic globe. Fairly common, even in the U.S. $25-$45.

BERKSHIRE - 1930s candy bar vendor. The base of this machine is chromium plated cast iron. Nice advertising decal on the cylinder glass globe. $300-$450.

BLOUNT, JAY WALTON - 1940s peanut vendor. This vendor, made of polished aluminum, is named after the vendor who created it for his own route use. $250-$350.

BLUEBIRD - Mid 1910s gumball vendor. These 1¢ vendors are constructed of aluminum. The machine comes with a special wrench, rather than a key, to open the top for refills. The machine came in several versions, with notably different bases. $250-$350.

BLUEBIRD HONEYDEW - Mid 1910s gumball vendor. Constructed of aluminum, this profit sharing machine has a large glass globe and a marquee bolted to the lid. These marquees, which explain how to redeem your prize, are a feature of most profit sharing machines. This vendor is also known as a 1-2-3 machine. The customer drops a penny in, pushes the handle to receive his gumball, and depending on luck, can push the handle a second and even third time to receive up to three gumballs for the same penny. Without marquee $350-$500; with marquee $650-$800.

BLUEBIRD - Late 1910s gumball vendor. This machine has profit sharing features similar to the Honeydew, but is designed with a tapered base. The machine pictured was purchased at a northern Wisconsin general store auction; it had resided there since the store's opening in 1922 until the auction in September of 2001. $400-$500.

BLOYD SANITARY CONFECTIONS -
Late 1930s peanut vendor. This 1¢
aluminum vendor is shown with a
replacement globe and top cap. The
original was manufactured with a
square globe. Most often found as a
peanut vendor, but I've also seen one
in a gumball version. $75-$125.

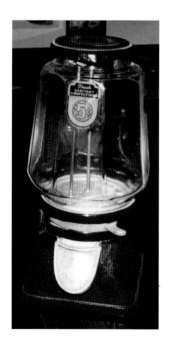

BLOYD LUCKY BOY - Late 1930s
peanut vendor. This machine has
aluminum construction and a unique
ten-sided globe. The machine pictured
was one of two dozen brand new Lucky
Boys found underneath the porch of a
house in Georgia. $75-$125.

BRICE WILLIAMS -1920s gumball vendor. This metal construction machine has a cast iron front, footed base, porcelain tray, and front and side glass windows. These machines are seldom found in working condition. $400-$700.

BUREL 1¢ BULK VENDOR - Late 1930s bulk vendor. Three compartment vendor made of steel construction. Difficult to find. $150-$250.

CADILLAC JUNIOR - 1950s peanut vendor. This small countertop aluminum vendor has a square plastic globe and came with an optional tray. While this can be a nice looking machine, it is often found in poor condition with a cracked globe or peanut damage. $50-$100.

CARLTON ROCKET - 1950s gumball vendor. This large plastic machine is shaped like a space ship. The price for this machine varies greatly depending on the quality of the plastic. A very popular machine, even with non-gumball machine collectors. $75-$175.

CLIMAX - early 1900s peanut vendor. This tall cast iron machine came in a couple of versions. Cast iron machine with unique ribbed globe (top right), rare; Hi-Lo version with horseshoe slug ejector (bottom left) $2500-$3000.

CLOVENA - Early 1900s breath pellet vendor. Wooden base with fancy glass globe. Rare.

COAST MULTI-VENDOR MACHINE
- Late 1950s gumball vendor. An
entertaining machine that came in
several sports versions, including
football, baseball, and basketball (not
pictured). When a penny is inserted,
the gumball shoots through the
playfield and into the dispensing
chute. Base is sheet metal with a
plastic globe. $150-$225.

COLUMBUS 21 - 1930s gum/peanut vendor. This machine is cast iron and comes with a barrel padlock and octagon glass globe. Designed for use in taverns, restaurants, and other establishments with limited space, it has several functional design variations, including standard gumball; gumball machine with attached ashtray; gumball machine with attached ashtrays, lamp shade and slug ejector. Standard model $250-$350; with ashtray $300-$400.

COLUMBUS 34 - 1930s gum/marble vendor. Available in cast iron or aluminum, this machine comes with an octagonal or round glass globe and a barrel padlock. The model shown at top left is a penny machine. The model shown at bottom right is a gambling version and takes a quarter. The customer puts his quarter in, gets a colored gumball, and matches it to the award card on or near the machine. Prizes are redeemed from the store clerk. Standard model $250-$450; gambling version, $300-$500 (add $40-$60 for an original Columbus barrel lock).

COLUMBUS 36 - 1910s tab gum vendor. This boxy cast iron machine could easily be mistaken for a match machine if not for the embossed Penny Gum writing on the front. $1200-$1500.

COLUMBUS 46 - 1940s gum vendor. This aluminum machine has an octagonal glass globe. Note the barrel padlock, also considered a valuable collectible. $150-$275 (add $40-$60 for an original Columbus barrel lock).

COLUMBUS MODEL A - 1920s gum/peanut vendor. This cast iron machine came with either an octagonal or round glass globe. Columbus A machines are distinguished by an embossed star on the glass globe and flap. $200-$325 (add $40-$60 for an original Columbus barrel lock). The model shown below is a Columbus machine made for the European/Belgium market. $275-$350.

COLUMBUS MODEL A - 1920s peanut/gum vendor. This cast iron machine came with either an octagonal or round glass globe. Columbus A machines are distinguished by an embossed star on the glass globe and flap. The three versions pictured here all have slug ejectors, making them extremely rare and valuable. The version shown below has a pac-man style slug ejector with a flat coin entry and tray built into the base. All rare.

COLUMBUS MODEL B -1910s gum/peanut vendor. This cast iron machine is similar in design to the Columbus Model A, but with a longer flap. Standard Model B $500-$700; Model B with tray built into base $700-$900; as shown here with optional slug ejector $850-$1250 (add $40-$60 for an original Columbus barrel lock).

COLUMBUS MODEL D - 1910s gum/peanut vendor. This cast iron penny machine came with a round glass globe, square base and legs. $600-$800.

COLUMBUS MODEL E -
1900s breath pellet vendor.
Small countertop machine
made of cast iron with a
unique round ribbed glass
globe. Rare.

COLUMBUS MODEL F -
1910s peanut vendor. Cast iron
machine produced with two
front feet and a round glass
globe. The decal states
"Proudly Worlds Best
Machine." $750-$950 (add
$40-$60 for an original
Columbus barrel lock).

COLUMBUS MODEL K -1910s
gum/peanut vendor. This cast iron
machine comes with a round glass
globe and has a slanted coin entry.
Columbus K machines are distinguished by an embossed star on the
flap. $450-$650 (add $40-$60 for an
original Columbus barrel lock).

COLUMBUS MODEL M - 1930s gum/peanut vendor. This machine came in either cast iron or chrome finish and with either an octagonal or round glass globe. Note the barrel padlock on the cast iron machine (left), also considered a valuable collectible. $150-$275 (add $40-$60 for an original Columbus barrel lock).

COLUMBUS MODEL V - 1930s marble vendor. Chrome plated machine with a unique cylinder glass globe. It has been used to vend marbles, gumballs, and even aspirin. $350-$500.

COLUMBUS DART - 1930s gumball vendor. Cast iron gambling machine came with a glass octagonal globe. The decal states "Watch the Dart, It's Fun, 1¢ Delicious Confection." When the customer enters his coin, the multi-colored dart at the inside/top of the machine spins. If the gumball matches the color of the dart, the customer wins a prize. Prizes are redeemed from the store clerk. $1200-$1600 (add $40-$60 for an original Columbus barrel lock).

CRYSTALET MACHINE - Early 1900s breath pellet vendor. This double vendor cast iron machine has fancy glass globes. Rare.

DEAN - 1970s peanut vendor. This Art Deco machine has aluminum construction and four glass panels for viewing the product. $40-$75.

DOUBLE NUGGET - 1930s peanut vendor. Two vendors in one. This machine is aluminum and can be found with either a glass or plastic globe. Plastic globe $75-$150; glass globe $125-$250.

EMPIRE - 1930s gumball vendor. The Empire State Building is embossed on the front of this cast iron machine. Glass front window. $250-$350.

ENTERPRISE - 1908 gumball/peanut vendor. The version pictured here is the earliest in a series of three. Tall cast iron base with glass globe. Rare.

FAUST MAIER - 1900s peanut vendor. This machine has a wood and metal base and a fancy light bulb shaped globe. Rare.

FEDERAL BREATH PELLET - 1900s breath pellet vendor. Small countertop machine with a cast iron base and cylinder shape globe. Rare.

FORD - 1930s gumball vendor. Ford gumball machines were made from 1919 to the 1960s. The machine has a metal base, sliding coin mechanism, and round glass globe. Some of the older Fords have an embossed logo on the glass globe. $100-$200 with reproduction globe; $400-$500 with original embossed logo globe.

FORD - 1950s gumball vendor. These gumball machines were produced with either a steel painted or chrome finish base. Globes came in either plastic or glass, and some had an optional marquee. These were often found on charity routes. $50-$90.

FORTUNE - 1930s gumball vendor. This machine had a tall, chrome polished metal base and glass jar style globe. Vended gumballs printed with your fortune. $600-$850.

FREEPORT - 1910s gum vendor. This machine came in several versions, all with embossed cast iron fronts and a wood body case. Rare.

FREEPORT DRAGON
- 1900s peanut vendor.
Two embossed dragons
decorate the front of
this heavy machine.
Cast iron front with
wooden sides and glass
window, optional side
view windows. Rare.

FREEPORT OWL - 1900s
peanut vendor. This
machine has wooden case
construction with ornate
cast iron embossing on the
front. Embossing reads
"Owl Vendor." Rare.

FREEPORT - 1900s peanut vendor. This wooden case machine has ornate cast iron scrolling and glass window front; it came with an optional marquee. A special feature of this machine is the small glass window above the chute that shows the last three coins inserted into the machine. Rare.

GRANDBOIS - late 1920s gumball vendor. Small countertop machine with cast iron construction and a glass cylinder globe. The manufacturer's name is embossed on the bottom of this machine. $150-$250.

GRISWOLD - 1910s peanut vendor. There were several variations of this machine manufactured throughout the decade. The base is footed cast iron with a large light bulb shaped glass globe. Pictured is the Red Star Griswold, with a red star embossed on the turn handle. $1700-$2300.

HANCE -1910s gumball/peanut vendor. Several versions of this desirable machine were produced. These machines have a cast iron base with a built-in tray and some were footed. Large egg shaped glass globe. From left: peanut version, rare; gumball version $1200-$1700; nickel peanut auto version with marquee, rare.

HANCE - REX VERSIONS - 1910s-1930s gumball/
peanut vendor. These durable old machines came in a
number of versions, including a hot nut and a breath
pellet machine. Top, left to right: standard hot nut
version $1800-$2400; hot nut with advertising top
$2000-$2600. Bottom, left to right: standard $1300-
$1700; gumball with cylinder globe $3200-$3800.

HANCE - REX VERSIONS - 1910s-1930s gumball/peanut vendor. These durable old machines came in a number of versions. Top, left to right: porcelain gumball with light bulb style globe and side dispenser $2800-$3500; cast iron gooseneck $2700-$3300. Bottom, left to right: breath pellet $2600-$3100; breath pellet w/solid glass globe rare.

HAPPY JAP - early 1900s chewing gum vendor. Made in the shape of a Japanese warrior's head, this cast iron machine worked using a clock mechanism. When the coin was dropped in, the clock mechanism pushed the gum out through his teeth. Rare.

HART - 1950s gumball vendor. This machine came with a chrome plated base and round globe and is sometimes mistaken for a Ford machine. $50-$100.

HAWKEYE - 1930s gumball vendor. This gambling vendor came in both aluminum and porcelain over cast iron versions with a glass globe. The machine had a bell built into its base. On the tenth pull, the bell would ring and the customer received his penny back along with the gumball. $150-$225.

HAWKEYE - late 1940s peanut vendor. Constructed of aluminum with a plastic globe, this more recent version of the Hawkeye was not nearly as nice as its predecessor in appearance and function. $50-$125.

HERSHEY BAR - 1930s candy bar vendor. Several versions of this slim wall mounted machine were available. The tall machine is constructed of sheet metal with a cast iron base. $75-$125.

HILO - 1900s peanut vendor. This machine was constructed of cast iron and has a nice football style glass globe. $1600-$2000.

HIT THE TARGET - 1950s penny drop gumball machine. A large, boxy gambling machine with an interesting space theme graphic on the front. The customer put his penny in the coin mechanism, turned the handle, and the penny dropped through the pin field. If he was lucky, he got his penny back along with the gumball. $75-$150.

HONEY BREATH BALLS - early 1900s breath pellet vendor. The four glass side windows on this aluminum machine allowed the customer to view the breath pellets as they moved through the glass tubes and into the chute. Nicely etched graphics on the window. Rare.

IDEAL - early 1930s peanut vendor. Aluminum construction with a fancy light bulb shaped glass globe. This machine featured a side coin entry and turn knob and came in several versions including hot nut. From left: standard aluminum version $600-$800; cast iron version, rare; hot nut version $1000-$1200.

JENNINGS IN THE BAG - mid 1930s peanut vendor. This machine has a large metal body and a small cylinder globe. Vended peanuts in small, wax paper bags. $250-$400.

JUNIOR PLAY - 1930s gum vendor. Made of metal with a glass front, this profit sharing mechanical vendor reads "A Game of Skill" on the front. Customers received two to six shots at the goal (depending on version) for 1¢ and were awarded a gumball for each shot made. On missed goals, the gum was recycled. Football version $2600-$3200; two-shot basketball (not pictured) $2800-$3300; three-shot basketball $3000-$3500.

LANGLEY'S - 1910s gum vendor. This small
countertop machine is made of pot metal with
a cylinder shaped glass globe. $800-$1200.

LAWRENCE - 1940s bulk peanut vendor. Double column vendor made of metal with front and side windows. $75-$150.

LEEBOLD - 1910s peanut vendor. Beautiful, ornate aluminum vendor with a glass globe. $2500-$3000.

LIBERTY - mid 1910s gumball vendor. Poorly constructed aluminum base with round glass globe. Rare.

LITTLE NUT B - 1930s peanut vendor. Small countertop vendor with elevated coin entry. Constructed of aluminum with glass globe. $100-$175.

LITTLE NUT C - 1930s gumball/peanut vendor. This small countertop vendor came in either cast iron or aluminum construction with a short, wide/round glass globe. Cast iron version $600-$800; aluminum version $500-$700. If you can locate this machine with the original embossed globe (a rare find) add $300.

LITTLE NUT HOUSE - 1930s peanut vendor. Cast iron countertop vendor shaped like a house. Glass windows. $1000-$1400.

LION - early 1900s gumball/ peanut vendor. This machine is constructed of cast iron with a light bulb shaped glass globe. Very ornate cast iron design with embossed coin entry, deep built-in tray, and cast iron feet with lions' heads. Painted graphics on side. $8000-$12,000. (Top and Bottom)

LOG CABIN - 1930s peanut vendor. This double bulk vendor constructed of polished aluminum can be found with either glass or plastic globes. $150-$275 (top and bottom).

MAGNA - 1930s peanut vendor. Constructed
of polished aluminum with a raised coin entry
and a sixteen-sided globe. $450-$750.

MANSFIELD - Early 1900s gum vendor. This hard to find machine is shown in good original condition. The mechanism is completely encased in glass, with ornate etched writing on the case. Missing marquee (as shown) $650-$1300; with reproduction marquee $850-$1500; with original marquee $1150-$1800.

MASTER - mid 1920s gum/ peanut vendor. Several versions of this machine exist. The square machines had a sheet metal body, porcelain finish top and base, and glass front and sides. Mechanism accepted either a penny or nickel. Ornate front coin entry. Gum/peanut version (top left) $150-$250; gooseneck version (bottom right) $350-$550.

MILLARD - 1910s breath pellet vendor. Small countertop machine came in a cast iron version as well as aluminum. Small glass cylinder shaped globe. Hard to find. $650-$850.

MILLS - 1930s stick gum vendor. This six column, chrome plated gum machine vended stick or pack gum. $75-$150.

MODERNE VENDOR - 1930s candy/gum vendor. There were several versions of this machine, including the small, cheaply made sheet metal countertop peanut version shown at left and the tall, wall mounted version that vended stick gum or candy bars. Both were constructed of sheet metal with a glass front window and cast iron base. $50-$100.

MO-JO - 1910s gum vendor. Two different manufacturers gave their vendor the same name. Chicle Products Co. made the metal base machine with the cylinder shape glass globe shown above and Roth & Langley made the cast iron machine with the domed glass globe pictured below. Chicle Mo-Jo $1400-$1700; Roth & Langley Mo-Jo $2000-$2500.

NATIONAL COLGAN'S TAFFY TOLU - 1900s gum vendor. This two column vendor is constructed of metal with a glass front and sides so you could view the product as it was dispensed. Nice graphics. $2600-$3200.

NATIONAL DISPENSER - late 1930s candy vendor. This four-way mint patty dispenser had a polished chrome finish and four glass product tubes. $150-$250.

NATIONAL SELF SERVICE - 1920s candy/gum vendor. This six column vendor is made of polished aluminum and has glass sides and front. $100-$200.

NATIONAL NOVELTY - 1910s gumball/
peanut vendor. Two versions of this cast
iron machine are pictured: a gumball
version with a round glass globe and a
peanut version with a light bulb shaped
globe. Gumball version $300-$500; peanut
version $450-$675.

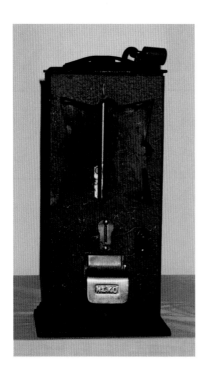

NEKO - 1930s peanut vendor. Both versions pictured here have metal and cast iron construction with a glass window. Single vendor $200-$350; double vendor $300-$500 (left and below).

NORTHWESTERN - 1910s gum vendor. This four-sided package gum dispenser is made of porcelain over cast iron with glass windows. Customers rotated the machine to their choice of packaged gum, dropped a coin in the top, and pushed the plunger. $1200-$1500.

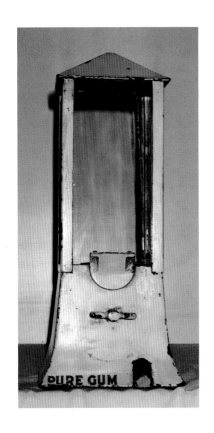

NORTHWESTERN - late 1910s gumball vendor. Square machine constructed of cast iron with four glass windows. The same model is pictured here in original condition and in restored condition. $500-$850.

NORTHWESTERN - mid 1920s gum vendor. This oddly shaped metal machine had glass in both the front and back. $400-$600.

NORTHWESTERN -1920s peanut vendor. This machine was produced in cast iron with a porcelain finish and glass globe. $2000-$2500.

NORTHWESTERN '31 MERCHANDISER - 1930s peanut vendor. There were two versions of this machine, including a standard peanut vendor and a penny drop gambling version. In the gambling version, the customer inserted his penny and it traveled through the playfield. If the coin dropped into the right slot the customer got his coin back with the gumball. Standard version $175-$275; penny drop version $800-$1200.

NORTHWESTERN '33 - 1930s peanut vendor. This machine was available in several colors of porcelain over cast iron and came with a standard or frosted globe. Some of the colors, including brown, blue, white, and yellow, are more desirable and add to the value of the machine. The most commonly found colors are red and green. The machine is also available in a penny drop gambling version similar to the Northwestern '31. Standard peanut version $150-$250.

NORTHWESTERN '33 - 1930s peanut vendor. This machine was available in several colors of porcelain over cast iron and came with a standard or frosted globe. Some of the colors, including brown, blue, white, and yellow, are more desirable and add to the value of the machine. The most commonly found colors are red and green. The machine is also available in a penny drop gambling version similar to the Northwestern '31. Standard peanut version $150-$250; penny drop version $800-$1200.

NORTHWESTERN '33 -
1930s gumball vendor. This
machine, like the peanut
version, can be found in
different colors of porcelain
finish over cast iron. Smaller
than the peanut vendor and
less flared in the base, this
machine is also more
difficult to find. $150-$175.

NORTHWESTERN '33 JUNIOR -
1930s peanut vendor. Small
countertop machine with cast iron
construction and an octagonal glass
globe. $450-$600.

NORTHWESTERN DELUXE - 1930s peanut vendor. The coin mechanism accepted either 1¢ for one turn of the handle or 5¢ for five turns. This tall machine was made in various colors of porcelain over metal finish and had either a glass or plastic globe. $125-$200.

NORTHWESTERN DUAL NUT - early 1950s hot peanut vendor. The top half of this large floor standing machine has a chrome finish with front glass viewing windows and is positioned on a metal stand. This double unit vendor also has a cup holder attached to the side. $250-$400.

NORTHWESTERN '39 - late 1930s peanut vendor. This cast iron machine was available in a standard version with different glass globes; short globe is pictured. Also available in a gambling gumball version called the '39 bell ringer. On the tenth pull, a bell (built into the base of the machine) would ring, drawing attention and creating excitement, and the customer would get his penny back along with the gumball. Standard peanut version $175-$275; gambling bell ringer version (below) $225-$325.

NORTHWESTERN '40 - early
1940s gumball vendor. This
machine was available in either
porcelain or painted steel.
Large glass globe. $125-$200.

NORTHWESTERN '49 - late 1940s
peanut vendor. Boxy steel construction with a square glass globe. The
whole front of this machine opens up
with the key, exposing the coin box
and allowing the route vendor to refill
the product. A very easy machine to
service and refill. $40-$75.

NORTHWESTERN TAB GUM - 1950s package gum vendor. This machine has a metal body, round plastic globe and top, and revolving display. You turn the top plastic piece to your selection before inserting a coin. $60-$100.

NORTHWESTERN JET - 1950s gumball/capsule vendor. This machine was available in either a chrome or painted steel version, with a plastic front display window. $50-$100.

NORTHWESTERN '60 - early 1960s gumball/peanut
vendor. This machine is made with a metal body and square
glass globe. Pictured are two machines mounted to a cast
iron stand. A common find, and a good starter machine.
$25-$75. Note: these machines are still being made today
in quarter or dime versions with plastic globes. $25-$50.

NORTHWESTERN Super '60 - early 1960s gumball/peanut vendor. Similar to the Northwestern '60 but with a larger capacity plastic globe. $25-$50.

NORTHWESTERN SATURN 2000 ROCKET SHIP - early 1960s gumball vendor. Shaped like a 1950s rocket ship with display windows, this clever machine came mounted on a cast iron stand. These vendors are popular with sci-fi collectors too. $350-$600.

NORTHWESTERN TRI-SELECTOR - mid 1930s bulk vendor. This three column vendor had a porcelain over metal finish with three square glass globes. The front had a selector dial, which could be turned to the desired product. The mechanism accepted either 1¢ for one turn or 5¢ for five turns. $400-$500.

OAK ACORN - late 1940s peanut vendor. This aluminum vendor comes with a square embossed glass globe. Oak Acorn machines are easily recognizable by the acorn decal and acorn embossed flap. This machine is still being produced today with slight variations. $25-$75.

OAK ACORN - 1950s gumball vendor. This
aluminum vendor comes with a square glass globe
and has an acorn embossed on the flap. $25-$75.

OAK ACORN - 1960s gumball vendor. This aluminum constructed machine has a square glass globe and the word "Oak" embossed on the turn handle. The machine pictured was re-painted in a red, white, and blue All-American theme. $25-$75.

OAK ACORN - late 1940s peanut vendor. This aluminum machine has a high capacity plastic globe. $25-$45.

OHIO MODEL 2 - 1930s gumball vendor. This machine has polished aluminum construction with a round glass globe. $350-$500.

OPERATORS ACE - 1930s gumball vendor. This aluminum construction machine has a unique glass globe. $1000-$1400.

PANSY - 1900s stick gum vendor. This nicely embossed double column vendor was made of either cast iron or aluminum. Customers received a stick of gum and their fortune for a penny. Rare.

PANTHER - 1950s gumball vendor. This machine has a slug ejector on the side (see yellow button) and embossed top cap and flap. Constructed of aluminum with a square glass globe. Very hard to find. $150-$250.

PENNY KING - 1930s gumball vendor. This machine has an aluminum base and a tall domed glass globe. Single $150-$250; double $275-$400.

PENNY KING 4-in-1 - 1930s peanut vendor. The customer turned this four compartment vendor to selected a product before inserting his coin. Note the fancy footed aluminum base and Art Deco styling. $750-$1100.

PERK-UP - 1940s breath pellet vendor. This aluminum construction machine came in several versions. The two shown feature two different globe styles. This vendor also came as a double unit on a tray. Single $60-$100; double $125-$225.

PIX SWEETMEAT - 1900s stick gum vendor. This machine, also called True Blue, has a metal base and a domed glass globe. The customer dropped his coin in the coin mechanism at the top of this two-column vendor and plunged the handle. Hard to find. $3200-$3700.

PULVER - 1899 candy/gum vendor. This tall, wall mounted vendor has a wood and metal case with front glass window. Nice graphics. The customer dropped his coin in and the puppet gave a nod of thanks. $2800-$3400.

PULVER - 1909 candy/gum vendor. This is the first tall case tin version Pulver. Nice painted lithographic front and sides. The customer dropped his coin in and the puppet gave a nod of thanks. A hard machine to find in good condition. $1800-$2200.

PULVER - 1910s gum vendor. This tall case porcelain machine has nice graphics and a glass paned front window. The customer dropped his coin in and the puppet gave a nod of thanks. $2000-$2500.

PULVER -1920s stick gum vendor. This short case wall mounted gum vendor has a porcelain finish and rounded corners. It also has a glass window with wire mesh to prevent the puppet and product from being stolen. Nice graphics on front. Some versions also had side graphics. Vendor with front graphics only (left) $650-$850; vendor with front and side graphics (bottom) $1200-$1400.

PULVER SELF SERVICE - 1930s gum vendor.
This five column bulk vendor was **not** coin-
operated. The customer selected his product
and pushed the plunger down, then paid the
store clerk for the product. $500-$700.

REGAL - 1930s peanut vendor. This machine came in aluminum or cast iron versions with a small cylinder globe. Cast iron (top left) $75-$150; aluminum (below) $75-$100.

REGAL - 1940s peanut vendor. This machine is made of aluminum. In the early 1940s the vendor came with a cylinder shaped glass globe. In the mid 1940s the design was changed to a pear shaped glass globe. $75-$125.

RO-BO - 1920s gumball vendor. This mechanical gumball machine was constructed of metal with a glass paned window. When the customer inserted his coin and pulled the handle, the gumball dropped down from the top display area. The figurine then scooped it up and dropped it down the chute. Rare.

RYEDE CHEWING GUM - 1920s stick gum vendor. This tall, slim wall mounted single column vendor was made of sheet metal with a mirror front. Notice the small product viewing window under the mirror. $275-$400.

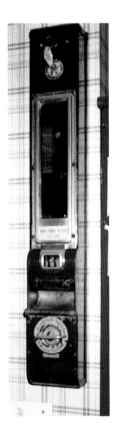

SCOOPY - 1950s gumball vendor. This mechanical machine was constructed of metal with a glass viewing window. The customer dropped his penny in, pushed the plunger, and the baker figure scooped a gumball from the oven and dropped it into the chute. A later version (bottom) shows a different cabinet design with the same mechanical function. First version $2500-$3000; later version $2200-$2500.

SELECTIVE FOUR WAY - 1930s bulk vendor. This heavy cast iron machine has a large four compartment glass globe. $400-$600.

SELL 'EM HOT - 1920s peanut vendor. This tall hot nut vendor was constructed of polished aluminum with a glass globe. The front red light tells you when the machine is on and this version has a side coin entry with a pull chain to release product. $250-$350.

113

SEL-MOR - late 1930s gumball/peanut
vendor. Made of cast iron with a tall
cylinder shaped glass globe. $175-$300.

SILVER KING - 1930s peanut vendor. This
penny vendor came with or without a
porcelain finish over cast iron and has a bell
shaped glass globe. Without porcelain finish
$100-$150; with porcelain finish $150-$225.

SILVER KING - 1940s peanut vendor. This machine came in a number of versions. All had aluminum construction with a glass globe and the words "TRY SOME" embossed on the flap. Pictured here is the standard version. $60-$100.

SILVER KING - 1940s peanut vendor. This machine came in a number of versions. All had aluminum construction with a glass globe and the words "TRY SOME" embossed on the flap. Pictured here are the the hot nut with ruby glass top and the ballerina version. The ballerina version provided entertainment along with the product. When the customer dropped a coin in, the ballerina figurine danced. Hot nut version $175-$250; ballerina version $400-$600.

SILVER KING - 1950s gumball vendor. Nicknamed the giant ace, this gambling gumball machine was made of aluminum with a large glass globe. This quarter machine was filled with some unusual color gumballs. The customer matched his gumball color to the display card to see what prize he had won. Prizes were awarded from the store clerk. $125-$175.

SIMMONS MODEL A - late 1930s peanut vendor. This machine is porcelain over cast iron with a fancy, square etched glass globe. $300-$400.

SIMPSON - 1910s peanut vendor. Hot nut vendor made of cast iron with a round glass globe. Notice the backwards "S" embossed on the flap. $1200-$1500.

117

SIMPSON - mid 1910s bulk vendor. The two versions pictured here, the Simpson Model A (top) and the slightly smaller Simpson Jr. (bottom) both had cast iron construction, round globes, and an embossed "S" on the flap. Model A $400-$600; Simpson Jr. $1000-$1200.

SIMPSON - 1920s gumball/peanut vendor. These machines were cast iron, chromium plated with a round glass globe. The profit sharing model shown below (with the nicely embossed marquee) is a combination machine in which the mechanism accepted either a penny or a nickel. Standard version peanut vendor $350-$450; profit sharing combination version $400-$650.

SIMPSON ARISTOCRAT - 1930s bulk vendor. This machine was made of chromium plated cast iron and brass, with a round glass globe and marquee. $375-$550.

SIMPSON DERBY- 1940s gumball/peanut vendor. This machine came in either a gumball or peanut version. It had aluminum or cast iron construction with a round glass globe. Customers dropped a penny in and the horses twirled like a carousel. $2400-$2800.

SMILIN' SAM - 1930s peanut vendor. Aluminum vendor in the shape of a man's head. The customer dropped a coin in and pulled out the tongue to get his product. Nice embossed lettering reads "SMILIN' SAM FROM ALABAM'." $3000-$3500.

SPECIALTY PETITE - 1930s breath pellet/peanut vendor. Constructed of aluminum with a lantern style glass globe, this small countertop machine came in a breath pellet or peanut vendor version. The customer dropped his coin in and pushed the extended button plunger. $900-$1200.

STA-HOT - 1930s peanut vendor. This hot nut machine has a tall, oddly shaped aluminum base with a large glass globe. Rare.

SNACKS - 1930s bulk vendor. Heavy triple column bulk vendor made of aluminum construction with glass front viewing windows. $100-$150.

STONER FRESH GUM - 1940s tab gum vendor. Six column wall mounted vendor constructed of sheet metal with a glass viewing window. $75-$150.

STRIKE IT RICH - 1950s gumball vendor. This gambling penny drop machine is similar to the machine called Hit The Target, also made in the '50s. Boxy steel construction with glass window to view the playing field. The customer entered his coin and if lucky, he got his penny back with the gum. $75-$150.

SUN - late 1940s peanut vendor. This machine has aluminum construction and glass windows on all sides. A common find, especially on the West Coast. $60-$100.

TOM THUMB - 1930s gumball vendor. Made of aluminum, this small countertop machine has a tall, slim glass globe and side coin entry. $250-$350.

TOY 'N JOY - 1950s capsule vendor. This boxy vendor was made to vend toy capsules. It is often found complete with the original display card taped to the inside front window as well as with leftover capsules. $10-$30.

UNIVERSAL ALMOND NUT -
late 1940s nut vendor. This
machine was private labeled for
Universal by Oak Mfg. Constructed of aluminum, the vendor
was available with either a plastic
or glass front and side windows.
The top cap is embossed with the
name Universal and this version
came with a map of the Americas
embossed on the flap. $25-$85.

UNIVERSAL - 1940s breath
pellet vendor. This machine was
private labeled for Universal by
Oak Mfg. Constructed of
aluminum with a standard glass
globe, the flap is embossed with
the Liberty Bell. $25-$85.

VENDEX FISHBOWL - 1930s gumball vendor. Aluminum base with fishbowl shaped glass globe and unusual fish decal. $125-$200.

VENDEX - 1930s gumball vendor. This cast iron base machine has a glass jar globe. $250-$350.

VICTOR - late 1930s gumball vendor. Several versions exist, all with cast iron construction and glass globes. Universal (top), one of the first Victor machines produced, $175-$275; Universal with tray built into the base (right) $250-$350; Topper (bottom) $100-$150.

VICTOR - 1940s gumball/peanut vendor. Several variations of this machine are shown. Top, left to right: the Victor Model V, which is constructed of cast iron with a round glass globe, $100-$150; the Half Cabinet peanut vendor, which was produced with either a glass globe or steel cabinet with plastic front viewing window (as shown), $50-$75. Bottom, left to right: Universal gumball vendor, which has a large aluminum constructed base with glass globe, $50-$100; and the Model K Sidewinder, $350-$450. The Sidewinder features a coin mechanism and turn handle on the side of the machine, and is made in a porcelain finish over cast iron with a round glass globe.

VICTOR - 1950s gumball vendor. Two versions of the victor are shown here. The Victor Half Cabinet (top left), which has aluminum construction with a plastic front viewing window, $35-$75; and the Topper (bottom right), which has aluminum construction with a glass globe, $40-$95.

VICTOR VENDORAMA - 1950s bulk
vendor. The peanut version (above) has
aluminum construction and a D-shaped
glass globe, $35-$75; the gumball/capsule
version (below) has aluminum construction and a plastic front view window with
an area for a card/prize display, $35-$60.

VICTOR BABY GRAND - 1950s gumball vendor. These wooden vendors had plastic view windows and came in a number of versions including (left) the Baby Grand Deluxe, $35-$75; and (below) the Five Star Baby Grand with stars embossed in the metal top, $50-$100.

VICTOR BABY GRAND - 1950s gumball vendor. These wooden vendors had plastic view windows and came in a number of versions including (left) the Five Star with card vendor attachment, $150-$250; and (below) the Jumbo 100 machine, which vended large size gumballs, $35-$75.

VICTOR BABY GRAND - 1950s gumball vendor. These wooden vendors had plastic view windows and came in a number of versions. This one is the Baby Granddad, a large machine that vended capsules. Note the small attached mirror on the top cap. $75-$150.

VICTOR - 1950s gumball vendor. These machines were wooden with a metal front and plastic front window and came in several sport theme versions. The customer dropped his coin in and used the flipper to play the game. This was for amusement only; the customer received one gumball, regardless of skill. Shown here are the Baseball and Football versions. $150-$250.

VICTORIAN SPLENDOR - early 1900s peanut vendor. This penny vendor is constructed of cast iron with a glass globe. Unusual shape. The customer dropped his coin in the top, turned the handle, and the peanuts dropped into the tray. Rare.

WAGNER - early 1900s gumball vendor. This polished aluminum machine dispensed small gumballs that traveled through the tubes, visible through the front and back glass windows. Rare.

WI-CO - early 1900s gum vendor. Heavy cast iron machine with glass windows. Rare.

WILBUR'S - 1920s chocolate bar vendor. This heavy metal four column vendor has a front mirror and display windows. $250-$400.

WILLIAM MICHAEL - 1930s gumball vendor. Cast iron construction with front feet and round glass globe. $350-$500.

YU-CHU - 1920s gumball vendor. Pot metal construction with a glass jar globe. $100-$160.

ZENO -1902 gum vendor. Wooden machine with nice graphics and front window. The machine pictured here is missing the mechanism. It was picked up in an antique shop outside of New Orleans where it was marked "Conversation Piece - $15." $700-$900 (for a complete machine).

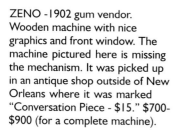

ZENO - mid 1910s gum vendor. This boxy wall mounted vendor was constructed of porcelain over steel. Small glass viewing window. $400-$900.

Cards

A.B.T. VENDING - 1930s card vendor.
These heavy, boxy metal machines were
used to vend sports, animal, movie star,
and baseball cards for 2¢ each. $175-$275

EXHIBIT SUPPLY - 1920s card vendor.
This small two column vendor was made
of metal and vended cards for a penny.
The graphic reads "Tom Mix - 32 Poses of
Your Favorite Cowboy." $150-$250.

OAK PREMIER - 1950s card vendor. This large
metal machine gave customers a gumball along
with the vended card. The machine came in
1¢, 2¢, 5¢, and 25¢ versions. $200-$300.

Matches

COLUMBUS 36 - late 1920s match vendor. This machine had a cast iron base and top with a sheet metal body. It vended either a book or box of matches for a penny. $175-$275.

COLUMBUS 48 - 1920s match vendor. This four column vendor was made of cast iron with a glass window front. Rare.

DIAMOND - mid 1910s match vendor. This unusual shaped machine has metal construction with painted lettering and vended books of matches for a penny. The clerk could set a dial within the machine (from 1 to 4) for how many books would be vended for the penny. The customer could then view this number on the top front of the machine before making his purchase. $400-$600.

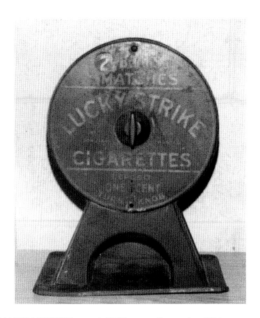

LUCKY STRIKE - mid 1910s match vendor. This unusually shaped metal construction machine has nice advertising graphics, and vended books of matches for a penny. The clerk could set a dial within the machine (from 1 to 4) for how many books would be vended for the penny. The customer could then view this number on the top front of the machine before making his purchase. $300-$400.

DIAMOND - late 1920s match vendor. This cast iron vendor has nice graphics on the front. The machine vended two books of matches for a penny. $150-$250.

EVERNICE - 1920s match vendor. This double column vendor was made of cast iron with a porcelain finish top and base and sheet metal body. It vended one box of matches for a penny. $150-$250.

HAWKEYE - 1930s match vendor. This vendor was constructed of sheet metal and vended two books of matches for a penny. $60-$105.

KNAPSACK - early 1900s match vendor. This octagon shaped machine was constructed of wood with three sides of viewing windows. Nice graphics. $2500-$3000.

MORREL - late 1920s match vendor. Constructed of aluminum and cast iron, this machine has two viewing windows that allowed the customer to see the mechanism move as well as vending the product. Rare.

NORTHWESTERN SELLEM - 1910s match vendor. Very ornate cast iron book match vendor came with two viewing windows, a cigar cutter, and match holder. $750-$1000.

NORTHWESTERN SCUP - 1910s match vendor. Cast iron book match vendor came with two viewing windows, embossed lettering, and nice graphics; it also has a cigar cutter and match holder built in. $1000-$1400.

NORTHWESTERN ART GRANITE - 1920s match vendor. This machine originally came in porcelain over cast iron (machine in photo was painted) and was made in two versions: two books for a nickel or one book for a penny. $150-$250.

NORTHWESTERN - 1920s match vendor. This machine came with a cast iron base and top and a sheet metal body. It vended one book of matches for a penny. Notice the unusual paint job on this one. $150-$200.

NORTHWESTERN BOOK MERCHANDISER - 1930s match vendor. This machine also has a cast iron base and top and a sheet metal body. It vended two books of matches for a penny. $50-$100.

OWL VENDOR - early 1900s match vendor. Constructed of a wooden base with cast iron feet and a domed glass globe, this double column vendor dispensed one box of matches for a penny. The cast iron marquee was embossed with "Eddy's Matches 1 Cent a Box." Rare.

RUSH HOUR - 1920s match vendor. This double column vendor has a cast iron base, sheet metal top and body, and four viewing slots. The customer dropped his penny in and the box of matches dropped into the base. The customer then opened a sliding metal door on the sides of the vendor to retrieve the product. $150-$200.

SAFETY MATCH - late 1900s match vendor. This four column vendor has a wood case body with ornate cast iron front and glass viewing window. It vended a book of matches for a penny. $300-$500.

SPECIALTY - 1910s match vendor. This machine came in two versions. The Advertiser version is ornate cast iron construction with a front marquee. The Perfection version is ornate cast iron with dragons embossed and a front glass viewing window. Hard to find. $1200-$1500.

Stamps

AMERICAN VENDING - mid 1910s stamp vendor. This double vendor had a wooden case and ornate cast iron front. Machine was produced in an All-American red, white, and blue theme. It vended two 2¢ stamps for a nickel or four 1¢ stamps for a nickel. $2200-$2600.

NATIONAL POSTAGE SERVICE - 1920s stamp vendor. This porcelain finish over metal machine has two display windows and a crank handle. $50-$75.

SCHERMACK PRODUCTS - 1930s stamp vendor. This machine has a cast iron base, metal front and back, and a crank handle. Front, top, and side glass windows allowed the customer to view the mechanism. It vended three 3¢ stamps for a dime. $150-$225.

SCHERMACK POSTAGE STATION - 1930s stamp vendor. Customers could rotate this chrome plated cast iron machine on a lazy Susan style base. The front side offered four 1¢ stamps for a nickel. Revolve the machine and the back side offered three 3¢ stamps for a dime. $150-$250.

SCHERMACK - 1940s stamp vendor. This small machine has a cast iron base and sheet metal body. It vended three 3¢ stamps for a dime or one 5¢ airmail and one 3¢ stamp for a dime. $30-$60.

SHIPMAN'S - 1930s stamp vendor. This machine has a porcelain finish cast iron base, chrome front, and sheet metal body. $150-$250.

SHIPMAN'S - 1950s stamp vendor. This boxy three column vendor was made with a porcelain front and sheet metal body. It has three small glass viewing windows. $40-$80.

Tobacco

BENNETT - 1900s cigar vendor. This case style vendor allowed the customer to serve himself. Using the tongs on top much like a claw vendor, the customer picked up his selection from the box of cigars and dragged it to the dispenser chute. The product was released down the chute after a coin was inserted. Rare.

ROI-TAN - 1950s cigar vendor. This large boxy metal vendor has a glass display window and porcelain front. We found two original boxes of cigars inside the machine when we purchased it. $175-$300.

GOLD LEAF - 1910s cigarette vendor. This large, boxy Canadian machine is made of wood and vended a small pack of cigarettes for a penny. $150-$200.

PENNY SERVICE - 1930s cigarette vendor. Small countertop machine made of sheet metal with a front glass window. It vended one cigarette for a penny. $275-$350.

ATLAS VAN LIGHT - 1930s lighter fluid dispenser. This lighter fluid dispenser has been converted into a lamp. The customer put a penny in the slot and pulled down the handle to dispense lighter fluid. $200-$300.

LIGHTER FLUID DISPENSER - 1930s lighter fluid dispenser. This countertop vendor was made of metal and glass and shaped like a gas station pump. The customer put a penny in the slot and pulled down the handle to dispense lighter fluid. A unique feature of this dispenser was that the store clerk could tell how much money was in the coin box by reading the gauge on the outside of the machine. $750-$950.

Miscellaneous

CINCH - 1950s vendor. Two versions of this wall mounted restroom machine are pictured. The machine on the right vends shoe shine cloths and the one on the left vended sanitary napkins. Sheet metal construction with nice graphics. $75-$125.

COIN-O-MATIC CASHIER -1940s coin changer machine. Art deco steel machine has nice embossed graphic. $125-$175.

HARMON - 1950s prophylactic vendor. This tall wall mounted machine was constructed of steel. $100-$175.

MASTER - 1920s prophylactic vendor. This tall wall mounted machine was chrome plated with painted graphics. $1200-$1600.

155

PARKING METER - 1950s parking meter. This parking meter was converted to a lamp. Parking meters are a common find at flea markets. $25-$50.

PERFUME VENDOR - 1920 perfume vendor. Metal construction with a large mirror on the front and nice decals. $500-$600.

VENROY - mid 1910s towel vendor. This cast iron machine has a wooden base and vended cloth towels. The display window has reinforced wire mesh under the glass. $300-$400.

ZENO and PRICE - early 1900s collar button vendors. The Zeno collar vendor (left) had a metal base and top, four glass sides, and a marquee (missing in the picture); it vended a collar button for a dime. The Price collar vendor (right) had a cast iron base and top, marquee, and cylinder globe; it vended a collar button for a nickel. Zeno version $400-$600; Price version $1100-$1400.

Foreign

AUTOMATEN - 1920s peanut vendor. This German machine is cast iron with a round glass globe. This machine will also take American quarters. $150-$300.

CHICLE - 1910s gum vendor. This Mexican wall mounted vendor was constructed of cast iron. Rare.

JUNIOR - 1920s peanut vendor. This German peanut vendor has an aluminum base and large round glass globe. $200-$300.

NATIONAL AUTOMATIC - 1930s candy/ cigarette vendor. This English machine was made in several versions to vend either candy or cigarettes. Metal construction. Products were both visible and secure behind a wire mesh reinforced glass viewing window. $150-$250.

NOBBY'S SALTED PEANUTS - 1920s peanut vendor. This large floor standing Australian peanut vendor is made of metal. Nice graphic on the front. Large glass globe with 3-D metal monkey on the lid. $2200-$2800.

Glossary

Here are a few terms you may encounter and their definitions

Bell Ringer. A small bell built into the base of the machine that would ring on every tenth turn, creating attention and excitement. You will typically find a bell ringer in Hawkeye and Northwestern '39 gambling machines.

Gambling Machine. A machine in which the mechanism was set to return the coin to the customer on the tenth turn, along with the gumball. Often used with a bell ringer. Some examples are the Hawkeye and Northwestern '39.

Profit Sharing. A machine in which the mechanism was set to vend variable amounts of product. The customer dropped his penny into the coin entry and turned the handle. Most times the machine allowed the standard one turn/gumball. But sometimes the machine allowed two or even three turns/gumballs for a penny. Some profit sharing machines, like the BlueBird 1-2-3, dispensed extra product based on pure luck. Others, such as the Junior Play basketball and football games, dispensed extra product based on the customer's skill level.

Hot Nut. When the store clerk plugs the machine in, a light bulb positioned at the top or bottom of the unit warms the product. This machine often comes with a red light on the top cap so you know when it's on.

Slug Ejector. Used to prevent the use of slugs. A valuable addition, slug ejectors are typically found on machines made from 1900 to 1940. Examples would be Advance and Climax vendors.

Barrel locks. A style of padlock often found on Columbus and other coin-op peanut/gumball machines produced in the 1930s. Barrel locks are collectible in and of themselves, and will add about $50 to the price of the machine.

Bibliography

Ayliffe, Jerry. *American Premium Guide to Jukeboxes and Slot Machines, 3rd Edition*. Books Americana, 1991.

Enes, Bill. *Silent Salesmen Too*. Walsworth Publishing Co, 1997.

Bueschel, Richard M. *Collector's Guide to Vintage Coin Machines*. Atglen, PA: Schiffer Publishing Ltd., 1997.